YOUR KNOWLEDGE HAS VALUE

- We will publish your bachelor's and master's thesis, essays and papers

- Your own eBook and book - sold worldwide in all relevant shops

- Earn money with each sale

Upload your text at www.GRIN.com
and publish for free

Bradley Tice

Random and Non-random Sequential Strings Using a Radix 5 Based System

GRIN Verlag

Bibliografische Information der Deutschen Nationalbibliothek:

Die Deutsche Bibliothek verzeichnet diese Publikation in der Deutschen National-
bibliografie; detaillierte bibliografische Daten sind im Internet über http://dnb.d-
nb.de/ abrufbar.

Imprint:

Copyright © 2012 GRIN Verlag GmbH
Druck und Bindung: Books on Demand GmbH, Norderstedt Germany
ISBN: 978-3-656-64526-9

This book at GRIN:

http://www.grin.com/en/e-book/199140/random-and-non-random-sequential-strings-
using-a-radix-5-based-system

GRIN - Your knowledge has value

Der GRIN Verlag publiziert seit 1998 wissenschaftliche Arbeiten von Studenten, Hochschullehrern und anderen Akademikern als eBook und gedrucktes Buch. Die Verlagswebsite www.grin.com ist die ideale Plattform zur Veröffentlichung von Hausarbeiten, Abschlussarbeiten, wissenschaftlichen Aufsätzen, Dissertationen und Fachbüchern.

Random and Non-random Sequential Strings Using a Radix 5 Based System

Bradley S. Tice
Advanced Human Design, P.O. Box 3868 Turlock, California 95381

Kolmogorov Complexity defines a random binary sequential string as being less patterned than a non-random binary sequential string. Accordingly, the non-random binary sequential string will retain the information about it's original length when compressed, where as the random binary sequential string will not retain such information. In introducing a radix 5 based system to a sequential string of both random and non-random series of strings using a radix 5, or quinary, based system. When a program is introduced to both random and non-random radix 5 based sequential strings that notes each similar subgroup of the sequential string as being a multiple of that specific character and affords a memory to that unit of information during compression, a sub-maximal measure of Kolmogorov Complexity results in the random radix 5 based sequential string. This differs from conventional knowledge of the random binary sequential string compression values.

PACS numbers: 89.70 Eg, 89.70 Hj, 89.75 Fb, 89.75 Kd

Traditional literature regarding compression values of a random binary sequential string have an equal measure to length that is not reducible from the original state[1]. Kolmogorov complexity states that a random sequential string is less patterned than a non-random sequential string and that information about the original length of the non-random string will be retained after compression [2]. Kolmogorov complexity is the result of the development of Algorithmic Information Theory that was discovered in the mid-1960's [3]. Algorithmic Information Theory is a sub-group of Information Theory that was developed by Shannon in 1948 [4].

1

Recent work by the author has introduced a radix 5 based system, or a quinary system, to both random and non-random sequential strings [5]. A patterned system of segments in a binary sequential string as represented by a series of 1's and 0's is rather a question of perception of subgroups within the string, rather than an innate quality of the string itself. While Algorithmic Information Theory has given a definition of patterned verses patternless in sequential strings as a measure of random verses non-random traits, the existing standard for this measure for Kolmogorov Complexity has some limits that can be redefined to form a new sub-maximal measure of Kolomogorov Complexity in sequential binary strings [6]. Traditional literature has a non-random binary sequential string as being such: [111000111000111] resulting in total character length of 15 with groups of 1's and 0's that are sub-grouped in units of threes. A random binary sequence of strings will look similar to this example: [110100111000010] resulting in a mixture of sub-groups that seem 'less patterned' than the non-random sample previously given.

Compression is the quality of a string to reduce from it's original length to a compressed value that still has the property of 'decompressing' to it's original size without the loss of the

2

information inherent in the original state before compression. This original information is the quantity of the strings original length before compression, bit length, as measured by the exact duplication of the 1's and 0's found in that original sequential string. The measure of the string's randomness is just a measure of the patterned quality found in the string.

The quality of 'memory' of the original pre-compressed state of the binary sequential string has to do with the quantity of the number of 1's and 0's in that string and the exact order of those digits in the original string are the measure of the ability to compress in the first place. Traditional literature has a non-random binary sequential string as being able to compress, while a random binary sequential string will not be able to compress. But if the measure of the number and order of digits in a binary sequence of strings is the sole factor for defining a random or non-random trait to a binary sequential string, then it is possible to 'reduce' a random binary sequential string by some measure of itself in the form of sub-groups. These sub-groups, while not being as uniform as a non-random sub-group of a binary sequential string, will nonetheless compress from the original state to one that has reduced the redundancy in the string by implementing a compression in each subgroup of the random binary sequential string. In other words, each sub-group of the random

3

binary sequential string will compress, retain the memory of that
pre-compression state, and then, when decompressed, produce the
original number and order to random binary sequential string.

The memory aspect to the random binary sequential string is, in
effect, the retaining of the number and order of the information
found in the original pre-compression state. This can be done by
assigning a relation to the subgroup that has a quality of
reducing and then returning to the original state that can be
done with the use of simple arithmetic. By assigning each sub-
group in the random binary sequential string with a value of the
multiplication of the amount found in that sub-group, a quantity
is given that can be retained for use in reducing and expanding
to the original size of that quantity and can be represented by a
single character that represents the total number of characters
found in that sub-group. This is the very nature of compression
and duplicates the process found in the non-random binary
sequential strings. As an example the random binary sequential
string [110001001101111] can be grouped into sub-groups as
follows: {11}, {000}, {1}, {00}, {11}, {0}, and {111} with each
sub-group bracketed into common families of like digits. An
expedient method to reduce this string would be to take similar
types and reduce to a single character that represented a
multiple of the exact number of characters found in that sub-

group. In this case taking the bracketed {11} and assign a multiple of 2 to a single character,1, and then reduced it to a single character in the bracket that is underlined to note the placement of the compression.

The compressed random binary sequential string would appear like this: [1000100101111] with the total character length of 13, exhibiting the loss of two characters due to the compression of the two similar sub-groups. De-compression would be the removal of the underlining of each character and the replacement of the 1's characters to each of the sub-groups that would constitute a 100% retention of the original character number and order to the random binary sequential string. This makes for a new measure of Kolmogorov Complexity in a random binary sequential string.

This same method of compression can be used with a radix 5 based system that provides for an even greater measure of reduction than is found in the binary sequential string. The radix 5 base number system has five separate characters that have no semantic meaning except not representing the other characters in the five character system. The following five numbers will represent the five characters found in the radix 5 base number system that will be used as an example in this paper: [0, 1, 2, 3 & 4]. As an example of a random radix 5 sequential string the following would appear like this: [001112233334440111223444]with a total character length of 24 characters. If all the applicable similar sequential 3 character's are compressed to a single representative character that represents the other two characters in the three character compressed unit of the string, then the following would result: [0012233334012234].

The underlined characters represent the initial position of the three character group of similar characters with a compressed state of 16 characters total. This is a reduction of one third the total original character length of 24 characters. A non-random radix 5 base sequential string will have the same character types: [0, 1, 2, 3 & 4] but with a regular pattern of groupings such as [00112233440011223344] that has a total character length of 20 and if all two sequentially similar characters are compressed using all 5 character types the following will occur: [0123401234] resulting in a compressed non-random radix 5 base sequential string of 10.

The paper has shown that a sub-maximal measure of Kolmogorov complexity exists that has implications to a new standard of the precise measure of randomness in both a radix 2 and a radix 5 based number systems.

References

[1]   S. Kotz and N.I. Johnson, Encyclopedia of Statistical
      Sciences (John Wiley & Sons, New York, 1982).

[2]   abide.

[3]   R.J. Solomonoff, Inf. & Cont. 7, 1-22 & 224-254 (1964), A.N.
      Kolmogorov, Pro. Inf. & Trans. 1, 1-7 (1965) and G.J.
      Chaitin, Jour. ACM 16, 145-159 (1969).

[4]   C.E. Shannon, Bell Labs. Tech. Jour. 27, 379-423 and 623-656
      (1948).

[5]   B.S. Tice "The use of a radix 5 base for transmission and
      storage of information", Poster for the Photonics West
      Conference, San Jose, California Wednesday January 23, 2008.

[6]   S. Kotz and N.I. Johnson, Encyclopedia of Statistical
      Sciences (John Wiley & Sons, New York, 1982).